I0805453

WOMEN INVENTORS

HIDDEN IN HISTORY

Petrice Custance

CRABTREE
Publishing Company
www.crabtreebooks.com

Author: Petrice Custance

Editorial director: Kathy Middleton

Editors: Sarah Eason, Ellen Rodger

Proofreaders: Jennifer Sanderson, Samantha Acker

Design: Paul Myerscough and Jessica Moon

Cover design: Emma DeBanks and Jessica Moon

Illustrations: Jessica Moon

Photo research: Rachel Blount

Production coordinator and Prepress technician: Margaret Amy Salter

Print coordinator: Katherine Berti

Written, developed, and produced by Calcium

Photo Credits:

t=Top, c=Center, b=Bottom, l= Left, r=Right

Inside: Department of Defense: Space and Missile Systems Center: p. 41; Shutterstock: 1000 Words Images: p. 13b; Africa Studio: p. 35; Alice-photo: p. 39; Anpannan: p. 26t; Farknot Architect: p. 12t; Blackday: p. 29b; Bolarzeal: p. 45; Eric Buermeyer: p. 6; CaseyMartin: p. 7t; Zhu Difeng: p. 38t; Everett Historical: pp. 21, 27, 32; FamVeld: p. 10b; Adam Jan Figel: p. 19; Gerald Robert Fischer: p. 9; Leszek Glasner: p. 4bg; Kaspars Grinvalds: p. 36; I_life: p. 13r; Kateryna Kon: p. 43; Morphart Creation: p. 11b; NicoElNino: p. 5t; Mariusz Niedzwiedzki: p. 37; Volodymyr Nik: p. 23b; Ohishiapply: p. 23t; Planner: p. 34l; Sihasakprachum: p. 17; Syda Productions: p. 25; TaraPatta: p. 24b; Tayvay: p. 44; Toa55: p. 15; Top Photo Corporation: p. 16b; Franco Volpato: p. 18l; Wikimedia Commons: p. 30t; African American Photographs assembled for 1900 Paris Exposition/Library of Congress: p. 28t; Cécile Didier: p. 10t; FA2010: p. 31t; Harris & Ewing/Library of Congress: p. 29t; Jim Henderson: p. 33t; Margareth E. Knight: pp. 24tl, 24tc; NARA: p. 7r; Nyttend: p. 33b; Post of Romania: p. 4; Jacques-Jean Tiziou: p. 42.

Library and Archives Canada Cataloguing in Publication

Title: Women inventors : hidden in history / Petrice Custance.
Names: Custance, Petrice, author.
Description: Series statement: Hidden history | Includes index.
Identifiers: Canadiana (print) 20200153226 | Canadiana (ebook) 20200153234 | ISBN 9780778772972 (hardcover) | ISBN 9780778773054 (softcover) | ISBN 9781427124760 (HTML)
Subjects: LCSH: Women inventors—Biography—Juvenile literature. | LCSH: Women inventors—History—Juvenile literature.
Classification: LCC T39 .C87 2020 | DDC j609.2/52—dc23

Library of Congress Cataloging-in-Publication Data

Names: Custance, Petrice, author.
Title: Women inventors hidden in history / Petrice Custance.
Description: New York, New York : Crabtree Publishing Company, 2020. | Series: Hidden history | Includes bibliographical references and index.
Identifiers: LCCN 2019054369 (print) | LCCN 2019054370 (ebook) | ISBN 9780778772972 (hardcover) | ISBN 9780778773054 (paperback) | ISBN 9781427124760 (ebook)
Subjects: LCSH: Women inventors--Biography--Juvenile literature. | Inventions--History--Juvenile literature.
Classification: LCC T36 .C87 2020 (print) | LCC T36 (ebook) | DDC 609.2/52--dc23
LC record available at https://lccn.loc.gov/2019054369
LC ebook record available at https://lccn.loc.gov/2019054370

Crabtree Publishing Company
www.crabtreebooks.com 1-800-387-7650

Printed in the U.S.A./022020/CG20200102

Copyright © **2020 CRABTREE PUBLISHING COMPANY**. All rights reserved. No part of this publication may be reproduced, stored in a retrieval system or be transmitted in any form or by any means, electronic, mechanical, photocopying, recording, or otherwise, without the prior written permission of Crabtree Publishing Company. In Canada: We acknowledge the financial support of the Government of Canada through the Canada Book Fund for our publishing activities.

Published in Canada
Crabtree Publishing
616 Welland Ave.
St. Catharines, Ontario
L2M 5V6

Published in the United States
Crabtree Publishing
PMB 59051
350 Fifth Avenue, 59th Floor
New York, New York 10118

Published in the United Kingdom
Crabtree Publishing
Maritime House
Basin Road North, Hove
BN41 1WR

Published in Australia
Crabtree Publishing
Unit 3 - 5 Currumbin Court
Capalaba
QLD 4157

Contents

CHAPTER 1

HISTORY'S HIDDEN HEROINES

Ask anyone to name a famous inventor and they would likely say Alexander Graham Bell, Thomas Edison, and Wilbur or Orville Wright. But what about female inventors? Most people would have difficulty naming just one female inventor, even though there have been many. Luckily, this is changing.

Inventing—Not Just for Men

Most people have heard of Alexander Graham Bell, the inventor of the first practical telephone, Thomas Edison, who invented the light bulb, and the Wright brothers, inventors of the airplane. There are countless books discussing the men who created such important inventions. But what about the inventor of the home security system? Or windshield wipers? Or the dishwasher? Many would describe those inventions as important, but the inventors of those technologies are largely unknown. Even lesser known is that those inventors were all women—and they certainly were not the only women inventors.

*Josephine Cochrane was tired of her china being chipped by her household staff as they washed the dishes. She decided to invent a machine to do a better job—and received a **patent** for her dishwasher in 1886.*

HIDDEN HISTORY

Brains Before Beauty

Hedy Lamarr was an Austrian-born Hollywood movie star of the 1930s and 1940s. Known at the time as "the most beautiful woman in the world," not many people knew she was also an inventor. During World War II (1939 to 1945), German boats were able to throw **Allied** weapons off course by jamming the **torpedo** signals. Alarmed by this, Hedy invented something called **frequency-hopping**, which kept the Germans from jamming a torpedo signal because they could not figure out where the signal would be. Hedy's invention is the basis for the **Wi-Fi** and **GPS** technology we use today.

Finding Hidden Heroines

There are many female inventors who should be better known. So many of the items and technologies we use every day were invented by women. Wi-Fi, coffee filters, and the foot-pedal trash can are just a few of them. Today, more and more research is being done to discover and share the fascinating stories of unknown or forgotten female inventors. The Women Inventors Network (WIN) is an organization dedicated to discovering and recognizing female inventors. WIN also works to encourage women and girls to study science and technology.

For a long time, many people did not believe Hedy had actually invented frequency-hopping. They thought such a beautiful woman could not be capable of achieving such a thing.

So, Why Can't We Read All About It?

The writers of history decide how someone is remembered after their life. They also decide who is remembered well beyond their lifetime. Up until recently, most of these writers have been men and most of them have focused their work on the successes of men. This means that many women's stories were never included in the history books.

Forgotten Heroines

Some women who achieved great things certainly received some praise in their day. However, when they were no longer alive to share their own stories and were overlooked by the writers of history, these heroines were forgotten. This is now changing as researchers and organizations are sharing the stories of women inventors and encouraging more girls and women to take up careers in the sciences.

Lost Einsteins

At Harvard University, researchers have come up with the term "lost Einsteins." This refers to the women, people of color, and poorer people who would have made important achievements, such as inventing, if they had had the opportunity.

Mary Anderson was a great mind whose windshield wiper invention is used by millions today.

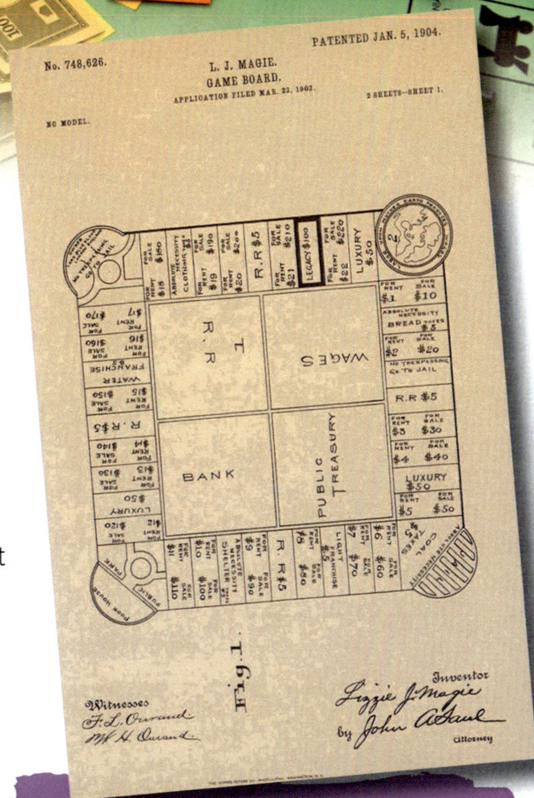

There is no way to know how many would-be Einsteins the world has lost, but efforts are now being made to give every woman inventor a chance. This is important because, for a long time, people believed women were not suited for or capable of scientific and creative thought. This led many who may have been interested in science and technology to give up and focus on more "suitable" work. That is why sharing the stories of history's heroines, and proving that women can and have contributed important inventions to the world, is so vital.

Cheated out of Her Due

There are several instances in history when women either did not receive praise for their invention or others tried to steal credit. For example, in 1903, when riding in a streetcar while it was snowing, Mary Anderson noticed the driver had to keep stopping to wipe the snow off the windshield. Mary wondered if a blade, operated from inside the streetcar, could be used instead. She drew up a plan for the windshield wiper and later that year, she received a patent. None of the car companies were interested in buying her invention but when Mary's patent ran out in 1920, they all began installing wipers.

In 1902, Elizabeth Magie Phillips created a board game called The Landlord's Game. In the 1930s, Charles Darrow presented a revised version of Elizabeth's game and claimed he invented it. His game, called Monopoly, made him a millionaire and, for a long time, Elizabeth was written out of the history of Monopoly.

CHAPTER 2

EPIC IN EUROPE

We have all heard of great male inventors from Europe who wowed people with their inspiring ideas. But working away behind the scenes were many amazing women whose inventions were just as life-changing as the men's. Let's meet some of them!

Jeanne Villepreux-Power, Heroine of the Oceans

It may be shocking to discover that the woman who invented the aquarium and became a highly respected scientific researcher had only a basic education—just enough to learn to read and write. But Jeanne Villepreux-Power had determination that made up for her lack of schooling.

From France to Italy

Jeanne Villepreux-Power was born in 1794 in a small village in France. At age 18, she walked 250 miles (400 km) from her home village to Paris to become a dressmaker. After just a few years, Jeanne was chosen to design the royal wedding gown for an Italian princess marrying into the French royal family. By being a royal dressmaker, Jeanne was introduced to many important people, including James Power, a wealthy English businessman. They married in 1818 and moved to Sicily, Italy.

Jeanne loved the sea and its amazing creatures, including argonauts like this one.

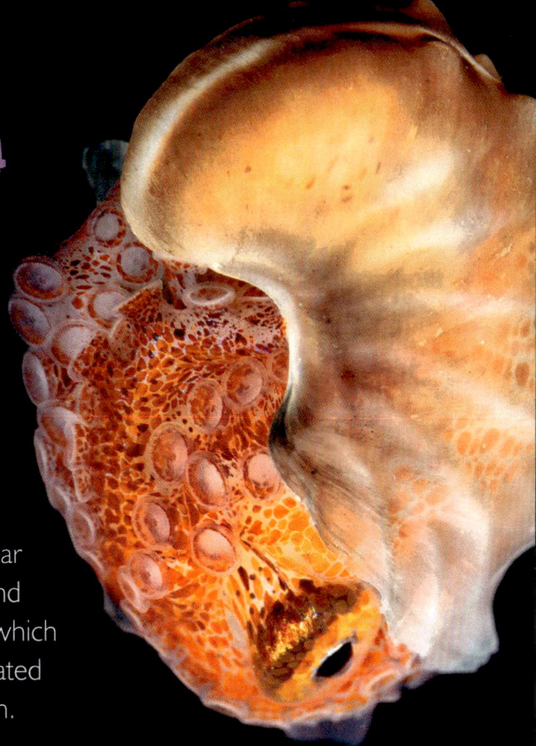

The soft body of an argonaut is protected by its tough shell.

Natural Wonder

While living in Italy, Jeanne became interested in plants and animals. She taught herself about the subject by reading, exploring the area around her, and collecting plants, **fossils**, and shells. Jeanne spent much time studying these samples and taking notes. She also became interested in **marine biology**. Jeanne lived near a bay, which was the perfect place to study the plants and animals that lived there. The bay had many argonauts, which are animals in the octopus family. Jeanne became fascinated by argonauts and focused most of her research on them.

Crawl into a Shell?

For centuries, scientists had been puzzled by the argonaut. They wondered how it developed its shell. By the 1800s, most scientists believed that argonauts found unused shells from other animals and then moved into them. But Jeanne began to suspect that argonauts actually made their own shells. She wanted to prove this idea but realized she would need to study live argonauts, not the dead ones she had found on the beach. So, in 1832, Jeanne became the first person to build an aquarium for the study of live water creatures.

If at First You Don't Succeed...

Jeanne's first aquarium was very similar to the fish tanks many people own today. She had several set up in her home and local fishermen helped her by carrying buckets of argonauts and seawater to her home and pouring them into the glass tanks. Unfortunately, the enclosed tanks were not suitable for the argonauts and most of them died. Jeanne realized she would have to keep the argonauts in an environment as similar as possible to the place where they are usually found, their natural **habitat**.

FIRST LADY

In 1832, Jeanne became the first female member of Italy's Catania Academy of Natural Sciences. She went on to become a member of several other scientific **societies**. In 1858, the highly respected scientist Richard Owen called Jeanne the "mother of aquariophily," or the study of life in tanks.

Power Cages

Jeanne next designed and built several wooden rectangular aquarium cages. The cages were made to sit in the water of the bay, with each corner weighted down by an anchor. Argonauts were caught and placed in the cages, which were nicknamed power cages. Being close to their natural habitat and surrounded by the moving bay water ensured the argonauts' survival. When Jeanne wanted to study the argonauts, the aquarium cages were raised out of the water.

Observations

Sur les mœurs de la Martre Commune

par

Madame Jeannette Power

née de Villepreux

Dédiés a madame la Comtesse de Santo Domingo

This is one of the research papers that Jeanne wrote when she was studying marine creatures.

Shell Shock

In her aquarium cages, Jeanne was able to observe some argonauts from their birth. She watched them develop their shells as they grew. Jeanne also slightly damaged the shells of some argonauts and observed them repairing their own shells. This proved that argonauts did make their own shells and that they did not move into leftover shells from other animals. Jeanne also discovered that the shells actually were not shells at all! They were egg cases, made to house a female argonaut's eggs until they hatched. This meant that all argonauts that had been studied, because they all had egg cases, were female. So where were all the males?

Thanks to Jeanne, today we can watch incredible sea creatures in aquariums such as this one.

HIDDEN HISTORY

Lost at Sea

Jeanne wrote about her findings, but at first many male scientists said that Jeanne's **claims** about the argonaut were impossible. But when faced with all the information she had gathered, they were forced to admit she was right. Jeanne continued researching with her aquariums. In 1843, she had all her notes and animal collection shipped to England, as she and her husband were moving there. Sadly, all her work was lost in a shipwreck. Deeply discouraged by the loss, Jeanne stopped researching.

Although she was not given the credit she deserved in her lifetime, today Jeanne is remembered for her important work. In 1997, a crater discovered on Venus was named Villepreux-Power for Jeanne.

Other scientists had found what they thought to be a wormlike creature living inside the egg cases, and believed that it was a **parasite** that lived off the female. Jeanne believed that the "worm" actually came from the male argonaut. She was later proven right by other researchers. In fact, while a female argonaut with her shell can measure up to 16 inches (40 cm), male argonauts are usually only 0.5 inches (1.25 cm) long. No one had noticed the males because they were so small! Male argonauts push the wormlike structure into the female egg case and then leave. Jeanne's aquarium cages allowed her to make some important discoveries. She also inspired other researchers to continue her work, and with her invention of the aquarium, she also changed how they work.

Forgotten and Found Again

After her death in 1871, Jeanne was largely forgotten, likely because of the loss of her research papers at sea. However, her writings and letters to other researchers from the time survive, and she is now being recognized for both her invention of the aquarium and for her groundbreaking research on argonauts.

Melitta Bentz, Cunning with Coffee

Melitta Bentz, born in Dresden, Germany, in 1873, loved to begin each day with a fresh cup of coffee. But she was tired of dealing with the coffee grinds that escaped the pot and floated in her cup. She was also tired of having to clean the coffee sludge out of the bottom of her pot every day. At the time, people boiled coffee grounds in pots. When they poured out the coffee, some gounds always escaped into the cup. Melitta decided there must be a better way to brew and enjoy her morning cup of coffee.

How many people would guess that the filters they use to make their daily brew were invented by a woman in the 1800s?

For the Love of Coffee

One day, Melitta ripped a piece of blotting paper, which is a thin, tissuelike paper, from her son's notebook. She poked some holes in the bottom of a pot with a hammer and a nail, placed the blotting paper inside, added some coffee, and then poured hot water over it. The coffee seeped through the paper filter, through the holes in the pot, and into a cup. But there were no grounds floating in it, and no mess to clean up.

Like many inventors, Melitta's idea was based on a desire to improve life—including a simple pleasure such as coffee-drinking.

A GOOD BOSS

Melitta cared very much about the people who worked for her. She introduced some unheard-of benefits at the time, such as a five-day work week, three weeks of vacation, and extra pay at Christmas time.

The Perfect Cup

In 1908, Melitta was granted a patent for the coffee filter. She and her husband Hugo opened a company, named Melitta, to sell the coffee filter. They displayed the coffee filter at the 1909 Leipzig Trade Fair and received interest from sellers across Germany. By 1914, the company had more than 12 workers and was producing thousands of coffee filters. It was also selling new-style coffeemakers to go with the filters. Melitta also hired people to go to stores and show customers how to use the product, which led to a jump in sales.

Success Story

Today, Melitta coffee filters are sold around the world. The company is still run by Melitta's family members and, in 2017, its sales totalled $1.8 billion. What Melitta accomplished was very rare at the time for a woman. It was also unusual that Melitta's husband was so supportive and willing to help his wife build the company.

Today, Melitta is a famous coffee filter brand with a name that is recognized worldwide.

Giuliana Tesoro, Making Life Safer and Easier

Whenever a firefighter puts on a uniform or someone unpacks a wrinkle-free shirt, they have Giuliana Tesoro to thank. Giuliana was born in 1921, in Venice, Italy. She was Jewish, and in 1938, she was denied entry to an Italian university because of **anti-Semitic** laws passed by the Italian government at that time. She moved to the United States and attended Yale University in Connecticut where she received a **PhD** in 1943.

Lifesaving Work

Giuliana worked for various chemical companies. Throughout her career, her inventions led to three main contributions to the making of textiles. In the 1970s, she invented a better way to make fabrics wrinkle-free by developing new chemical coatings. She also invented coatings that prevented **static** in textiles. Both of those inventions were popular, especially with travelers. However, Giuliana's most significant invention was flame-retardant fabrics. These fabrics were either fireproof or significantly slower to burn. This made a huge impact on uniforms and clothing worn by emergency response workers.

Guiliana's fascination with fabrics would change the future of our world and make it a much safer place for everyone.

Giuliana's flame-retardant fabric has made the job of firefighting much safer than it was before her invention.

Flame-retardant fabrics also became standard for curtains and furniture in schools, hotels, and theaters. In cases of fire in public spaces, flame-retardant fabrics help slow down the spread of flames, giving people time to reach safety.

A BUSY LADY

When she was not inventing, Giuliana was teaching. In 1972, she became a visiting professor at the Massachusetts Institute of Technology (MIT) and remained there until 1982. She then worked as a professor at Polytechnic Institute of New York University until she stopped working in 1996.

A True Inspiration

Giuliana received more than 125 patents throughout her career, many of them involving flame-retardant fabrics. At the time, she was one of the most patented women in the world. Giuliana Tesoro is an excellent example of a person fleeing **persecution** who found safety in North America and went on to contribute greatly to the world.

CHAPTER 3

AWESOME IN ASIA

Asia has produced some amazing women inventors, from heroines who invented lifesaving cures for diseases to women who made martial arts a thing of beauty. Let's meet some of these incredible inventors from the East.

Ng Mui, Elegant Yet Fierce Fighter

Put on any **martial arts** movie and chances are you will see actors providing the **kung fu** action. But here is the kicker: many of their moves were invented by a woman. Her name was Ng Mui and she is credited with inventing several styles of martial arts. One of the most popular today is Wing Chun, a form of southern Chinese kung fu and a style of self-defense.

Few people realize that the inventor of the most popular type of kung fu was Ng Mui.

Bird Watching

In the 1700s, Ng Mui was a **Buddhist nun** of the Shaolin Temple near Zhengzhou, China. It is believed that one day, Ng Mui watched a crane and a snake fight. Ng Mui admired the way the crane used its legs and wings to both fight and defend itself at the same time. The crane's moves were graceful and controlled, and it never lost its balance. Soon after, Ng Mui put those moves into practice.

People who practice martial arts try to move with the grace and beauty of a crane.

Wing Chun relies on skill and fitness rather than bulk and strength. Even small, slim people become powerful when using it.

STEEL MAGNOLIA

Wing Chun means "eternal springtime" in English. It is known as an elegant form of fighting and does not require physical strength or a heavy weight for a fighter to be effective. This makes it a great martial art form for everyone, especially women.

First Student

Ng Mui wanted to help a young girl, who a village bully was intent on marrying, and used this style of kung fu to defeat him. The new style could be learned quickly and did not require great strength. It depended on quick arm movements, learning an opponent's weaknesses, and using instinct to figure out the opponent's next move. The girl, Yim Wing Chun, used this style of kung fu to defeat the bully. Today, the style is known as Wing Chun.

On the Big Screen

Wing Chun kung fu has been shown in several action movies. Generations of fans have enjoyed watching Wing Chun practiced by greats such as Bruce Lee and Jackie Chan. The story behind Wing Chun has also been explored on screen. In 1994, Malaysian actress Michelle Yeoh played Yim Wing Chun in a movie. In a 2010 movie based on her life, Ng Mui is played by Chinese actress Kara Wai.

Dr. Flossie Wong-Staal, Providing Hope Through Science

In the early 1980s, the disease now known as **AIDS** first appeared. AIDS attacks a person's **immune system**, leaving them unable to fight off infections. By 2018, more than 35 million people had died from AIDS. Although there is currently no cure, great strides have been made in treating this disease, and AIDS patients are now living longer and healthier lives. Much credit for the successes in AIDS research must go to the scientist and inventor Dr. Flossie Wong-Staal.

Studying Science

Flossie was born in 1947 in China. In 1952, her family moved to Hong Kong. In school, Flossie was encouraged to study science, which she did not like at first, but soon came to love. In 1965, Flossie moved to the United States to study at the University of California, Los Angeles. After receiving her PhD in **molecular biology** in 1972, Flossie began working at the National Cancer Institute in Bethesda, Maryland.

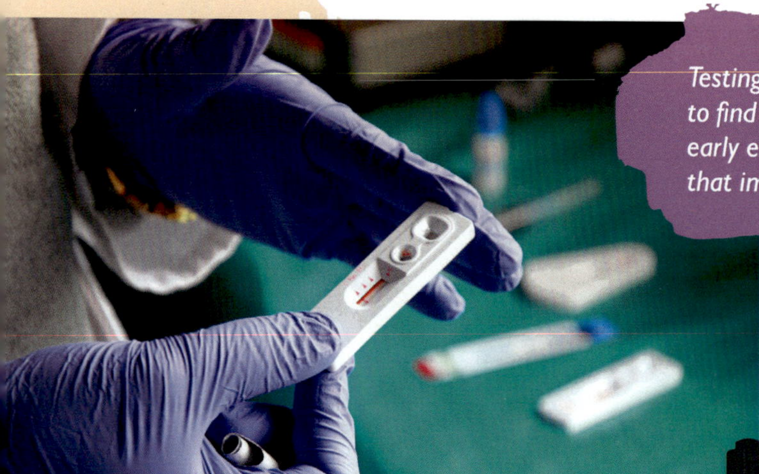

Testing for AIDS has made it possible to find out if people have the disease early enough to receive treatment that improves their lives.

Flossie was fascinated by ***viruses*** *and how they work, spending many hours studying them under microscopes in the lab.*

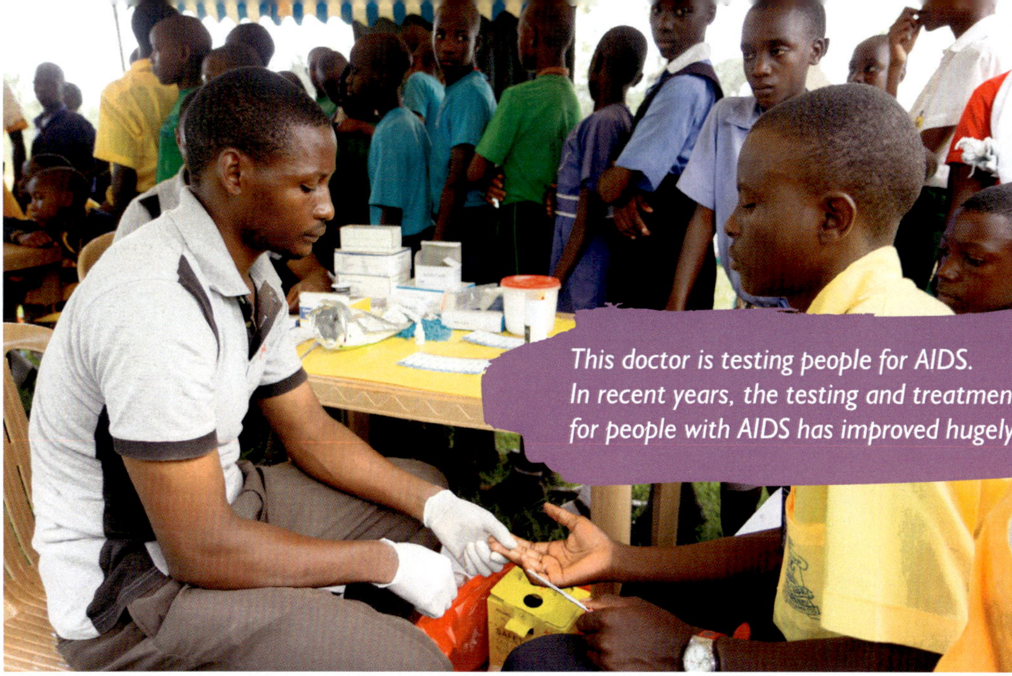

This doctor is testing people for AIDS. In recent years, the testing and treatment for people with AIDS has improved hugely.

The Map Is the Key

In 1983, Flossie became one of the first scientists to discover the **human immunodeficiency virus (HIV)**. The next year, she cloned the virus. This means that she made an exact copy of the virus. This was a huge breakthrough, because the cloning process allowed researchers to create a map of the virus's **genes**. That map helped Flossie and her colleagues prove that HIV was the cause of AIDS. From this knowledge, Flossie invented the first HIV tests. Her research also led to the effective AIDS treatments that exist today. Flossie holds patents for several **protocols** and techniques related to HIV and AIDS testing. On September 14, 2019, she was **inducted** to the National Women's Hall of Fame, putting her on track to receive the widespread recognition she has long deserved.

TYPHOON FLOSSIE

Flossie was born with the name Yee Ching Wong. The teachers at her Hong Kong school encouraged Yee's parents to give her an English name. This was likely because Hong Kong was under British rule at the time. Her father chose the name Flossie after a typhoon (a tropical storm that forms in the northeast Pacific Ocean) that had recently struck the area. It may seem odd to name a daughter after a storm, but now it seems quite appropriate. AIDS hit the world like a deadly storm and Dr. Flossie Wong-Staal challenged that storm. She helped calm the waters and gave HIV and AIDS patients hope through her important research and inventions.

CHAPTER 4

AMAZING IN THE AMERICAS

The Americas are famous for their fast-thinking inventors. When you think about these lands, you'll likely think of many amazing inventions, but I bet you've never thought that the life raft, the paper bag, and the foot-pedal trash can were invented by women! Let's find out just how they did it.

Margaret Knight, Paper Bag Queen

Today, Margaret Knight is recognized as the inventor of the flat-bottomed paper bag—but only because she fought for it. She was born in 1838 in York, Maine. From the time of early childhood, Margaret loved to work with tools and build toys such as kites and sleds. Margaret received only a basic education, so she taught herself how to make things.

Hard Worker

Margaret's father died when she was young. At 12 years old, she left school to work at a cotton mill, where cotton is made into yarn, to support her family. The conditions at the mill were very bad because there were no laws to protect the safety of workers, especially children. Margaret worked at the mill 12 to 14 hours a day, six days a week.

*Margaret loved working with her hands (see box opposite), so working as an **engraver** must have been an enjoyable job for the young inventor.*

Necessity Is the Mother of Invention

During her first year working at the mill, one of Margaret's coworkers was badly hurt when a metal piece flew off one of the **looms**. Margaret was very upset by the accident and she worked to find a solution. She invented a metal guard to hold the shuttles—the casing for the yarn—in place. The mill owner quickly began using Margaret's metal guards, and other mills in the area also began using them. Margaret was still very young, barely 13 years old, and had no idea that she could apply for a patent, which would allow her to make money from her invention. But she never made that mistake again.

AHEAD OF HER TIME

Margaret stopped working at the cotton mill in her late teens as it was harming her health. Over the next 10 years, she worked at many jobs, including home repair, photography, **upholstering**, and engraving–none of which were common jobs for women. Margaret enjoyed all of those jobs as they gave her a chance to work with many different tools.

In a cotton mill, workers would spin or weave cotton into yarn or cloth. The mills were extremely hot, loud, and filled with cotton dust, endangering workers' health.

The Big Idea

In 1867, Margaret started working at the Columbia Paper Bag Company in Springfield, Massachusetts. Her job was to fold and bundle paper bags. At the time, paper bags had envelope-like openings. Flat-bottomed paper bags, which are much better for holding items such as groceries, were not common at that time because they were so expensive to make. Margaret began to wonder if a machine could be made to produce the more user-friendly paper bags.

Brown-Bagging It

Margaret began researching and drawing up designs for a flat-bottomed paper-bag-making machine. At the time, she wrote in her journal, "I am told that there is no such machine known as a square-bottomed machine. I mean to try away at it until I get my ideas worked out." Within a year, Margaret had invented a machine to cut, fold, and glue flat-bottomed paper bags. Her wooden machine was 2.5 feet (76 cm) long and 1 foot (30 cm) wide. During testing, Margaret's machine produced 1,000 bags.

Margaret was practical and she could see the value in flat-bottomed paper bags, which were much easier for people to carry their groceries in.

HIDDEN HISTORY

Whose Honor Is It?

Surprisingly, there is one honor that Margaret didn't rightfully earn—and it probably belongs to another of history's hidden heroines. In Framingham, Massachusetts, a sign was placed on the home in which Margaret lived in, declaring her the first U.S. woman to receive a patent. But Margaret was not the first. That honor likely belongs to Mary Kies, who received a patent in 1809 for inventing a method of weaving straw with silk.

Ironing It Out

Margaret knew her flat-bottomed bags would be popular with the public. So, determined not to make the same mistake as with her invention at the cotton mill, she decided to apply for a patent. Margaret's model was made of wood, but she learned that it would need to be made out of metal to apply for the patent. She had an iron model made and then went to apply for her patent. However, she was shocked to discover that someone else had already put forward her plans and got the patent.

Margaret's flat-bottomed paper bag machine was a brilliant invention. It helped Margaret turn her cool idea into a money-making business.

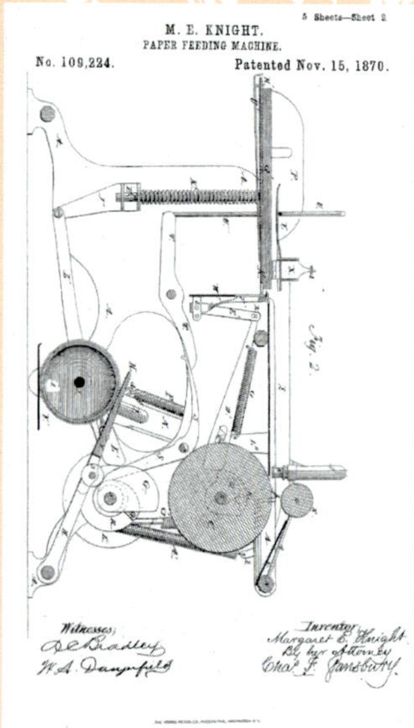

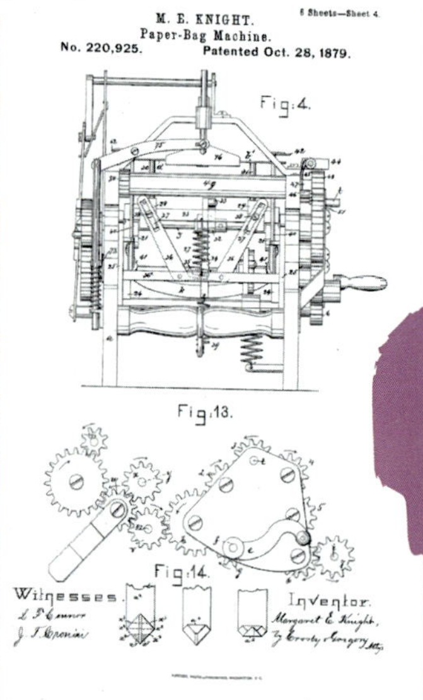

In court, there was no stopping Margaret and no one could argue that she was not the inventor of her paper-bag machine when they saw her sketches (at left)!

Nice Try

While the iron version of Margaret's paper-bag-making machine was being made, a man named Charles Annan spotted it. He stole the plans and applied for the patent himself. Margaret hired a lawyer and took Annan to court. In the courtroom, he claimed that it was impossible that a woman could have come up with the idea, and insisted the idea for the machine was his. Margaret presented her very detailed notes, sketches, and plans as proof that she had done the work to create the machine. The court sided with Margaret and she received her patent in 1871.

When a steel-tipped yarn shuttle, such as this one, fired out of a loom and injured a worker, Margaret quickly came up with a safety device to keep it from happening again.

Business Woman

Margaret opened the Eastern Paper Bag Company. In the beginning, she had some difficulty with the male workers at her company. They could not believe that a woman could understand machinery and run her own company. These problems did not last, and the company was soon quite successful. Department stores and grocery stores began buying her bags.

An Inventing Machine

Over her lifetime, Margaret received more than 20 patents. Some of her other inventions include a a machine for precision-cutting of leather, lid-removing pliers, and improvements to **combustion engines**. In a 1913 article on women inventors in *The New York Times*, the writer noted that Margaret, aged 75, was working 20 hours a day on her 89th invention. Despite this, she never became wealthy. When Margaret died, she had only $300. Today, paper bags are still made by machines based on Margaret's design.

Truly a Genius

Margaret Knight is finally becoming recognized for the genius inventor that she was. Several books have recently been published about her. In 2006, she was inducted into the Paper Industry International Hall of Fame. Her original paper-bag-making machine can be viewed at the Smithsonian Museum in Washington, D.C.

LADY EDISON?

Margaret passed away in 1914. In a tribute to her after her death, a newspaper referred to her as a "Lady Edison." This was likely meant as a compliment at the time, but it suggests that a woman inventor is not a normal thing, while a male inventor, such as Thomas Edison, is normal. Margaret herself once said, "I'm only sorry I couldn't have had as good a chance as a boy." Thankfully, the world has come a long way. And no one is calling female inventors "Lady Edisons" anymore.

The next time you use a paper bag in a store, think of Margaret and her amazingly inventive mind!

Maria Beasley, Making Us Safer at Sea

Maria Beasley was born in Philadelphia in 1847. Before Maria became an inventor, she was a dressmaker. Not much is known about her life or how she became an inventor, but we do know that she became a wealthy woman because of her inventions. Surprisingly, the 1880 U.S. **census** lists her as an unemployed housewife.

Saving Lives

It is not known how Maria became interested in improving life rafts, but it's not hard to understand why. Before her work, emergency life rafts were just wooden planks strapped together. Maria's goal was to invent a raft that was "fire-proof, compact, safe, and readily-launched." Maria's new and improved life raft had guard rails and rectangular metal floats. The raft also folded and unfolded easily for use and storage. Maria received a patent for her life raft in 1882. The life rafts were used on the *Titanic*, the ship that hit an iceberg and sank on its first voyage. Maria's life rafts saved more than 120 lives on the *Titanic* alone.

Thanks to Maria, many of the people who survived the Titanic disaster did so because of the life rafts on board.

Pay Day

Maria is credited with 15 other inventions, including a wooden barrel-making machine, a steam generator, and a device to keep trains on their tracks. Her wooden barrel-making machine earned her $20,000 a year, which is about $450,000 today.

HIDDEN HISTORY

Tell Your Story

Maria Beasley is an example of a great woman who has been forgotten by history. If she had written her own story, she would have had more recognition. Perhaps she is a lesson that women must not be shy about sharing their stories and detailing their achievements. We do know about one instance when Maria Beasley was not shy. In 1884, she proudly exhibited several of her inventions, including the modern life raft, at the Cotton Centennial Exposition in New Orleans, Louisiana.

Madam C.J. Walker, Millionaire Mom

Sarah Breedlove Walker was born in 1867 in Louisiana. Her parents Owen and Minerva were recently freed slaves. Sarah became America's first African American female millionaire and moved into the same neighborhood as John D. Rockefeller, one of the richest men in the United States.

In the early 1900s, it was unthinkable that a woman could become a millionaire, let alone an African American woman such as Sarah, pictured above.

Hair Trouble

Before her success, Sarah, later known as Madam C.J. Walker, experienced hardship. She became an orphan at 7, a wife at 14, a mother at 17, and a widow at 20. Unable to read, Sarah worked as a laundress to support her daughter. Then, one day, her hair began to fall out.

Sarah had been suffering from dandruff and other scalp problems. This could have been because most haircare products available at that time were not appropriate for African American hair. Sarah began to learn about proper haircare from her brothers, who were barbers. Then, at the 1904 St. Louis World's Fair, Sarah met Annie Malone, an African American businesswoman selling haircare products. Sarah began selling Annie Malone's products but was already planning a larger role for herself.

Sarah was a smart businesswoman—she knew that women would pay for good products suited to their hair.

MOVING ON UP

Sarah was deeply inspired by African American leader Booker T. Washington's 1901 book *Up From Slavery*. In his book, Washington encouraged African Americans to lift themselves up from the effects of racism through education, self-reliance, and never giving up.

Sarah lived up to Booker T. Washington's (shown right) ideas of not giving in to racism but instead overcoming it through hard work and smart thinking.

A New Path

Sarah took all that she had learned from Annie Malone and invented her own line of haircare products for African Americans. Sarah's special blend included substances such as petroleum jelly and sulfur. She called her products the Wonderful Hair Grower. She also invented the Walker System, a process of scalp preparation, lotion, and the use of an iron comb.

Sarah's special products for caring for African American hair fed into a huge market—women all over the United States began to use her haircare line.

Sarah wanted to make a good impression wherever she went. She drove a smart-looking car and was always beautifully dressed.

So Very French

In 1906, Sarah married Charles Joseph Walker, who was brilliant at advertising. Sarah was now known as Madam C.J. Walker. She added the French "Madam" to her name as she thought it sounded elegant and created the right image for her company. With her husband helping to advertise the products, Sarah's sales skyrocketed. Her line grew to nearly 20 products, including cosmetics such as skin-care products. Advertisments were found in African American newspapers and magazines.

Sarah knew how to win people over with words and was captivating when speaking in public about her products.

The Personal Touch

Sarah went on the road to promote her business. She personally went door-to-door selling her products. In churches and other meeting places, Sarah showed how to use her haircare system. She understood that her image was very important to sales, so she took public-speaking lessons. Sarah became so successful that her company expanded into Cuba, Haiti, Costa Rica, and Jamaica. In 1910, Sarah opened her company head office in Indianapolis. It included a factory, salon, beauty school, and a lab for research.

Madam C.J. Walker's products were a hit wherever she went, and women lined up to get their hands on her miracle-working lotions.

Beauty Army

Sarah began training a national network of sales agents for door-to-door selling. She called them her "beauty culturists," and they wore a uniform of white shirts and black skirts. She soon became the largest employer of African American women. Sarah was a strong believer in women making their own money, so all her agents had the opportunity to earn money based on their sales. She also provided them with the skills to open their own businesses.

THE SECRET

Sarah once said, "There is no royal flower-strewn path to success. And if there is, I have not found it. For if I have accomplished anything in life it is because I have been willing to work hard."

Giving Back

As her wealth increased, Sarah became more and more devoted to charity work and helping others. She gave generously to African American schools, old-age homes, orphanages, or homes for parentless children, and The National Association for the Advancement of Colored People (NAACP). Sarah also began to take on more **activism** work. She started giving talks on issues affecting African Americans, such as racism.

Stand for Something

In 1917, a mob of white people murdered around 100 African Americans during riots in Illinois. Sarah led a protest group and traveled to Washington to campaign for new laws to protect African Americans.

SHE MEANS BUSINESS

In 1912, Sarah spoke at the National Negro Business League (NNBL) convention. The organization, started by Booker T. Washington in 1900, was created to support the advancement in business of African Americans. The organization still exists today as the National Business League. At the convention, Sarah said that although she started life in the cotton fields of the South, she worked hard to be promoted, first to a laundress, then to a cook, and finally, to making hair goods and preparations, which were made in her own factory on land she owned.

Sarah went from selling her products door-to-door to owning her own factory, laboratory, and beauty school, where she trained sales agents.

Sarah also encouraged her beauty culturists to stand up for their beliefs. That year, she held a conference in Philadelphia for her workers. It was possibly the first time in the United States that women had gathered to discuss business. At the 1917 Philadelphia conference, Sarah encouraged her sales agents to do their duty and speak against the racist policies of the government.

Sarah not only dressed to impress, but she also lived in one of the most impressive houses in New York City, shown above.

Home Sweet Home

Sarah had become well known for her love of luxury. In 1917, she asked Vertner Tandy, the first African American architect in New York City, to design a wonderful home for her. At the time, it cost $250,000 to build. The home was built in the same neighborhood as some of the richest white men in the country. Sarah was born the daughter of enslaved people, yet made such a success of herself that she moved into the same neighborhood as a Rockefeller!

In 1927, this building in Indianapolis was created to be the headquarters for Sarah's company. Today, it is an historic landmark.

Lillian Gilbreth, the Kitchen Helper

Some people find it funny that the woman who dedicated her life to improving efficiency in the kitchen, and who invented several helpful kitchen gadgets, could not cook. When she once did attempt to cook, her children referred to the creation as "dog's vomit on toast." But Lillian Gilbreth had much more important things to do than cooking!

Home Lab

Born in 1878, Lillian was one of the first female engineers to earn a PhD. She and her husband Frank both worked as experts in ergonomics, which is the study of how equipment in the workplace should best be placed for the worker's comfort and efficiency. Lillian and Frank had 13 children, and they were able to watch their children washing dishes and brushing their teeth to discover more efficient ways to do such everyday tasks.

As a mom of many children, Lillian knew that women often had their hands full! That is why she invented the foot-pedal trash can, which allowed people to open the can with the touch of a foot.

Whisking food by hand is hard work and takes a long time. Lillian knew there was an easier and quicker way to do it—cue her electric mixer!

LIFE ON THE BIG SCREEN

Both the 1950 and 2003 film versions of *Cheaper by the Dozen* are based on the 1948 book written by two of Lillian's children. The story is about the adventures of the Gilbreth children living with two efficiency experts for parents.

Everything and the Kitchen Sink

Lillian invented many helpful kitchen appliances, including the foot-pedal trash can, electric mixer, shelves inside fridge doors, and an improved electric can opener. She also advised the General Electric company on the ideal modern kitchen layout, including the height of counters and sinks. In recognition of her work, Lillian was given a place in the National Women's Hall of Fame in 1995.

Imagine a life without shelves inside your fridge door! Thanks to Lillian, we have efficient fridges that help make our everyday lives a lot easier.

Marie Van Brittan Brown, Making Us All Safer

In the 1960s, Marie Van Brittan Brown was a nurse in Queens, New York City. Marie and her husband Albert both worked during the day and sometimes at night, so she was often home alone and had begun to feel nervous in her apartment. The crime rate was increasing in Marie's African American neighborhood and the police often took a long time to respond to emergencies. The problem of mostly white police forces being slow to respond to calls in African American neighborhoods was common in the 1960s, during the time of the **civil rights movement**. Marie decided to take matters into her own hands and found a way to feel safe in her own home.

As a nurse, Marie spent much of her time thinking about how to help people. Making sure they were safe in their homes was a natural next step.

Feeling safe at home is important to people. Marie's invention jump-started a trend in home security that has continued to this day.

LIKE MOTHER LIKE DAUGHTER

Marie and Albert's daughter Norma followed in her mother's footsteps. She too became a nurse and an inventor. Norma has several patents for inventions related to health issues and protecting women from violence.

Many buildings are fitted with security systems today, such as call systems through which an occupant can speak to the person outside their home.

First of Its Kind

With the help of Albert, who was an electrician, Marie invented the very first home security system. Four peepholes at different heights were drilled into her door. A cabinet containing a movable camera was then attached to the door. The camera would run up and down over the peepholes, recording whoever was on the other side of the door. The camera was connected to a screen in Marie's bedroom. From there, she was able to view who was at the door and even speak to them through a two-way microphone. The security system also included the ability to lock and unlock the door from a distance, as well as a panic button that would immediately contact the police or signal neighbors for help.

Thank You, Marie

In 1969, *The New York Times* published an article about Marie and Albert and their invention. Appearing in the national newspaper made her and her invention well-known. In the article, Marie discussed her hopes to make and sell their home security system to other people.

Marie would have been inspired by the technology that people now have to help them feel safe in their homes, such as a video-link to callers at the door.

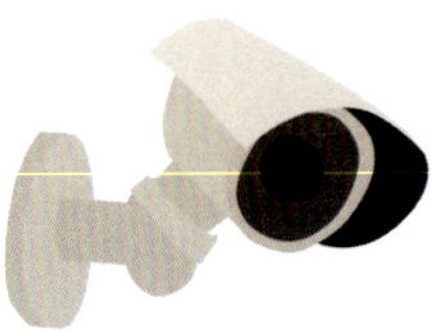

Lady First

Marie and her husband applied for a patent for their home-security system in 1966, and received it in 1969. On the patent, Marie's name appears first, proving that while Albert certainly provided his technical skills to the inventing process, the concept for the home-security system came from Marie.

Big Brother

Today, the television monitoring part of Marie's system is known as closed-circuit television (CCTV). A version of CCTV was first used by the Germans during World War II (1939 to 1945), but Marie was the first to apply CCTV technology for use in people's homes. She set the stage for all modern home-security systems and the video-monitoring systems that are now used in public spaces all around the world.

Do you think Marie would have approved of all uses of CCTV today? How do you think she would imagine the technology being used, and for what reasons?

While helpful in allowing people to feel safe and for recognizing criminals, the use of CCTV in public spaces has many people worried about rights. Some worry that governments and people in positions of power are able to use CCTV technology in illegal ways, such as monitoring people with different political views.

What Happened?

Home-security systems are a billion-dollar business. But it does not appear that Marie ever attempted to make money from her invention. Perhaps she and Albert sold the patent and did make money, but there are no records to suggest this. Several inventors who filed patents after the 1960s for similar technology list Marie's patent as an inspiration for their work.

HIDDEN HISTORY

Recognized, Then Forgotten

Marie Van Brittan Brown is the perfect example of an inventor known in her lifetime, written about in a national newspaper, but still forgotten by history. After *The New York Times* article, there are no other records of press coverage about Marie. Very little is known about the rest of her life, other than she passed away in 1999.

CCTV has its place in keeping people safe in public and in private. It can help police solve crimes and allow people to watch what is going on outside their homes.

Yvonne Brill, Powering Communications

Yvonne Brill, born near Winnipeg, Manitoba, in 1924, wanted to study science. Her high school science teacher told her that a woman would never get far in science, and her father encouraged her to forget science and open a store instead. Yvonne ignored all of them. She graduated from the University of Manitoba in 1945 and moved to the United States to continue her studies at the University of Southern California. At that time, it is believed Yvonne was the only woman in North America studying rocket science.

It IS Rocket Science

In 1967, Yvonne invented a **propulsion** system. This system gives satellites the **thrust** to be able to fly in space on their own. Yvonne patented her propulsion system for satellites in 1972. In 1983, the first communications satellite using her invention was launched. Yvonne's invention is still used today by satellites handling worldwide phone service and some television broadcasts.

Yvonne was a forward-thinking woman who followed her childhood dreams of working in science. She never gave up on her goals and went on to change the world through her inventions.

Can you think of all the things you do every day that rely on satellites? How many things can you list?

In 2011, President Barack Obama awarded Yvonne with the National Medal of Technology and Innovation. It is the nation's highest honor for engineers and inventors. Yvonne is an example to all young inventors that gritty determination can help you reach your goals.

HIDDEN HISTORY

Hiding Her Genius

After Yvonne passed away in 2013, *The New York Times* published a tribute to her that began, "She made a mean beef stroganoff, followed her husband from job to job and took eight years off from work to raise three children." Many people couldn't believe such an important newspaper would start an article on such an important woman like that. After the backlash, *The New York Times* rewrote the beginning of the article to read, "She was a brilliant rocket scientist who followed her husband from job to job and took eight years off from work to raise three children."

CHAPTER 5

ACING IT IN AFRICA

Women in Africa have a long history of inventing things that have changed the lives of people for the better. Today, many more African women are making the most of new opportunities to make their mark in the field of science and technology. Let's meet some of the women whose groundbreaking inventions represent the exciting future of Africa.

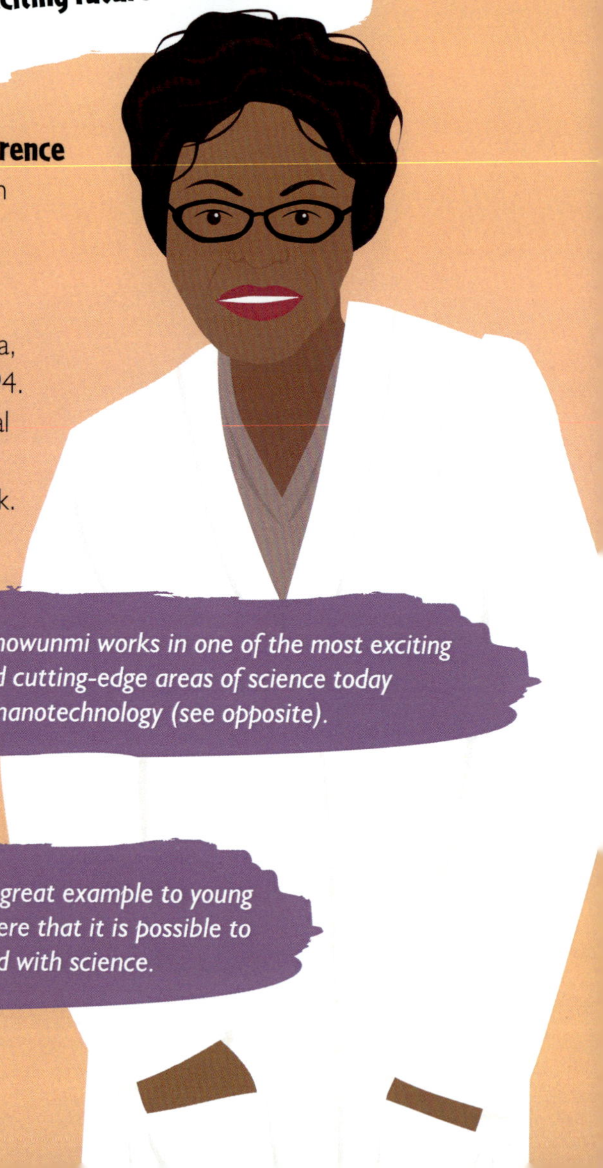

Omowunmi Sadik, Support Makes All the Difference

Omowunmi Sadik was born in Lagos, Nigeria, in 1964, into a family of scientists that encouraged her to also explore her world through science. She studied at the University of Lagos and then attended Wollongong University in Australia, where she received her PhD in chemistry in 1994. She was a researcher for the U.S. Environmental Protection Agency (EPA) before becoming a professor at Binghamton University, in New York.

Omowunmi works in one of the most exciting and cutting-edge areas of science today —nanotechnology (see opposite).

Omowunmi is a great example to young women everywhere that it is possible to change the world with science.

This is an artist's impression of a tiny surgical machine at work inside a body. Scientists believe that this type of nanotechnology will one day be possible.

Life-changing Inventions

Omowunmi holds five U.S. patents on biosensors. Biosensors are devices that sense chemicals. Omowunmi's inventions in the area of biosensors can be used in incredible ways, from picking up cancer early to pain management in patients and even the detection of drugs and explosives for law enforcement. She has also worked on technology to remove and recycle dangerous chemicals found in certain types of waste, such as from factories.

Small Scale, Big Impact

In 2012, Omowunmi cofounded the Sustainable Nanotechnology Organization (SNO). Nanotechnology is an area of science that looks at the control of **matter** on a tiny scale. SNO is researching and spreading awareness about the exciting possibilities of nanotechnology, such as making new medicines and producing clean electricity.

MANY HONORS

In 2003, Omowunmi was the first Nigerian to receive Harvard University's Distinguished Radcliffe Fellowship Award. In 2016, she received the Nigerian National Order of Merit Award, which is granted to Nigerians who have made outstanding contributions to the nation.

Jessica Matthews, Powering the Future

Jessica Matthews is a Nigerian American who is bringing clean energy to the world in a very special way. In 2008, while a student at Harvard University, she came up with an idea for an engineering assignment. She remembered that when attending her aunt's wedding in Nigeria, the power had gone out. Diesel generators were used as a power backup but they had produced dangerous fumes. Jessica thought there must be a safer way to create power.

Getting a Kick Out of It

For that engineering class, Jessica and her classmate Julia Silverman presented Soccket, a soccer ball that produced clean power by harnessing kinetic, or movement, energy. For every 30 minutes the ball was played with, up to 3 hours of clean, cost-efficient power was generated.

Jessica is inspired by the idea of using science to improve life for people in developing countries.

Jessica's Soccket was a great invention and shows how engineering can lead to new ideas that can change the world.

Jessica founded the company Uncharted Power to sell Soccket, with the idea of providing electricity where it can be scarce, such as in parts of Africa. Providing people with an easy, clean, and cheap way to produce power can be lifechanging. Jessica also developed a skipping rope that produced energy in the same way as Soccket.

HIDDEN HISTORY

A Long Way to Go

In an interview, Jessica discussed being a woman in the tech world. She said that in the tech industry, men are seen as better than women, and many companies expect men to be in positions of power, not women. She feels that it takes people a long time to understand and believe in her as a scientist because she is a woman. Jessica is working hard to change these attitudes in her field.

Shining a Light

Jessica is determined to help the more than 1 billion people around the world who do not have access to reliable power. She has recently partnered with Disney to develop new ways to produce, transfer, and store electricity in those communities. Her close links to Nigeria mean that Jessica is particularly eager to improve life for people in that country. She is also working on building a **hydroelectric** dam in Nigeria and is invested in creating practical solutions that will benefit the Nigerian people.

Jessica is working hard to create improved hydroelectric dams that will help provide cleaner energy for the people of Nigeria.

GLOSSARY

activism The raising of awareness to bring about political or social change

AIDS An acronym for the Acquired Immunodeficiency Syndrome, which is a disease that weakens a person's immune system

Allied The nations, including England, the United States, and Canada, which fought against Germany in World War II

anti-Semitic Describes a person or something that shows prejudice against Jewish people

Buddhist nun A female teacher of the Buddhist religion, which follows the writings of Buddha

census An official counting of a certain population of people

civil rights movement The activism of African Americans in the United States, from the mid-1950s to the mid-1960s, to gain equal rights under the law

claims Statements about something

combustion engines Engines that burn fuel in order to run, such as a car's gasoline engine

engraver A person who cuts or engraves words or a design onto the surface of an object

fossils The remains of dead plants and animals that lived a long time ago

frequency-hopping When radio signals randomly jump to different channels while they are being sent

genes The parts of cells in living things that determine how the living things will grow and develop

GPS An acronym for Global Positioning System, a satellite system used to determine an exact position on Earth

habitat The place in which a plant or organism is usually found

human immunodeficiency virus (HIV) A virus that attacks a living thing's cells and can lead to the development of AIDS

hydroelectric Related to electric energy created by using water power

immune system The internal part of a living thing's body that helps it fight off infection

inducted When a person is officially added to an organization

kung fu A particular style of martial art developed in China, often used for self-defense

looms Devices used for the weaving of yarn to make fabric

marine biology The study of living things found in oceans

martial arts Traditional forms of hand-to-hand fighting without the use of weapons

matter The substances that make up a physical object

molecular biology The study of how different parts of living cells work together

parasite An organism that lives on or in another organism and feeds off it

patent An official document giving an inventor ownership rights to their invention

persecution The cruel and harmful treatment of a person or group of people

PhD The highest degree in any discipline at a university or college

propulsion The act of pushing something forward

protocols Official rules on how something should be done

societies Groups of people who meet to discuss an interest, such as art or music

static The build-up of an electrical charge on objects, often caused by objects rubbing together

thrust The force that pushes something in a certain direction

torpedo An underwater missile fired by a ship or submarine

upholstering Covering of furniture in fabric

viruses Small particles that cause disease and can spread from one organism to another

Wi-Fi The technology that allows devices, such as smartphones, to wirelessly connect to the Internet

LEARNING MORE

Read more about women who broke the rules and changed the world!

Books

Adams, Julia. *Women Who Made History: Inventors and Scientists*. Gareth Stevens Publishing, 2019.

Jones, Grace. *People Who Changed the World: Science and Arts* (Ideas, Inventions, and Innovators). Crabtree Publishing, 2019.

Sullivan, Otha Richard. *Black Stars: African American Women Scientists and Inventors*. Wiley, 2012.

Thimmesh, Catherine. *Girls Think of Everything: Stories of Ingenious Inventions by Women*. HMH Books for Young Readers, 2018.

Websites

Visit this site to discover more incredible stories about women inventors:
www.amightygirl.com/blog?p=12223

This site will inspire you to put your inventor's hat on!
http://inventivekids.com

You'll find inventing challenges, cool games, and fun design videos at:
https://pbskids.org/designsquad

At the United States Patent and Trademark Office kids' page, you'll meet young inventors who are changing the world:
www.uspto.gov/kids/inventors-kids.html

INDEX

ABOUT THE AUTHOR

Petrice Custance has written books on many fascinating topics, but discovering and writing about the amazing women inventors who have changed the world has been one of her most exciting projects ever! Through this book, Petrice hopes to inspire other young female inventors to get into technology and create their own amazing inventions.